New York City Bank Building Fire:
Compartmentation vs. Sprinklers
New York, New York

Investigated by: J. Gordon Routley

This is Report 071 of the Major Fires Investigation Project conducted by TriData Corporation under contract EMW-90-C-3338 to the United States Fire Administration, Federal Emergency Management Agency.

Department of Homeland Security
United States Fire Administration
National Fire Data Center

U.S. Fire Administration Fire Investigations Program

The U.S. Fire Administration develops reports on selected major fires throughout the country. The fires usually involve multiple deaths or a large loss of property. But the primary criterion for deciding to do a report is whether it will result in significant "lessons learned." In some cases these lessons bring to light new knowledge about fire--the effect of building construction or contents, human behavior in fire, etc. In other cases, the lessons are not new but are serious enough to highlight once again, with yet another fire tragedy report. In some cases, special reports are developed to discuss events, drills, or new technologies which are of interest to the fire service.

The reports are sent to fire magazines and are distributed at National and Regional fire meetings. The International Association of Fire Chiefs assists the USFA in disseminating the findings throughout the fire service. On a continuing basis the reports are available on request from the USFA; announcements of their availability are published widely in fire journals and newsletters.

This body of work provides detailed information on the nature of the fire problem for policymakers who must decide on allocations of resources between fire and other pressing problems, and within the fire service to improve codes and code enforcement, training, public fire education, building technology, and other related areas.

The Fire Administration, which has no regulatory authority, sends an experienced fire investigator into a community after a major incident only after having conferred with the local fire authorities to insure that the assistance and presence of the USFA would be supportive and would in no way interfere with any review of the incident they are themselves conducting. The intent is not to arrive during the event or even immediately after, but rather after the dust settles, so that a complete and objective review of all the important aspects of the incident can be made. Local authorities review the USFA's report while it is in draft. The USFA investigator or team is available to local authorities should they wish to request technical assistance for their own investigation.

This report and its recommendations were developed by USFA staff and by TriData Corporation, Arlington, Virginia, its staff and consultants, who are under contract to assist the USFA in carrying out the Fire Reports Program.

The USFA greatly appreciates the cooperation received from the Fire Department of the City of New York, with special thanks to Deputy Chief Steven C. DeRosa, 3rd Division; Deputy Chief Herbert (Ted) Rohlfing, Fire Prevention, Special Projects; and Supv. Fire Marshal Tom Clark for the information and assistance they provided.

For additional copies of this report write to the U.S. Fire Administration, 16825 South Seton Avenue, Emmitsburg, Maryland 21727. The report is available on the USFA Web site at http://www.usfa.dhs.gov/

U.S. Fire Administration
Mission Statement

As an entity of the Department of Homeland Security, the mission of the USFA is to reduce life and economic losses due to fire and related emergencies, through leadership, advocacy, coordination, and support. We serve the Nation independently, in coordination with other Federal agencies, and in partnership with fire protection and emergency service communities. With a commitment to excellence, we provide public education, training, technology, and data initiatives.

TABLE OF CONTENTS

New York City Bank Building Fire:
Compartmentation vs. Sprinklers
January 31, 1993

Local Contacts: Deputy Chief Steven C. DeRosa, 3rd Division
Deputy Chief Herbert (Ted) Rohlfing
Fire Prevention, Special Projects
Supv. Fire Marshal Tom Clark
Fire Department of the City of New York
250 Livingston Street
Brooklyn, New York 11201-5884
(718) 403-1354

OVERVIEW

A highrise office complex in the heart of midtown Manhattan was the scene of a major fire on the night of January 31, 1993. It was the most destructive highrise fire in New York City in more than a decade, resulting in direct property damage of more than $10 million and a much larger loss due to business interruption and secondary effects.

The fire, which originated on the sixth floor, spread to the seventh and was extending into the eighth floor before it was controlled. These floors are at the maximum reach of outside aerial equipment and elevated master streams, which were used successfully to control upward propagation of the fire. If the fire had originated at or above the tenth floor, it would have been much more difficult, if not impossible, to stop the successive involvement of higher floors, even with the rapid response of more than 400 fire suppression personnel.

This fire is particularly significant as an evaluation of the effectiveness of Local Law 5, the retroactive requirements that were enacted for all highrise office buildings in New York City, after a series of destructive fires in the 1960s and 1970s. It suggested that the compartmentation option offered by Local Law 5 may be inadequate to prevent fires from growing to extremely destructive proportions and placing both occupants and firefighters in danger of death or injuries. Although the extent of the fire was not as great as similar fires that occurred in Los Angeles and Philadelphia in recent years, it could have equaled or exceeded their magnitude if it had originated on a higher floor. This result reinforces the opinion of many fire experts and authorities having jurisdiction that automatic sprinklers should be required in all existing highrise buildings, as well as new construction.

1

SUMMARY OF KEY ISSUES

Issues	Comments
Compartmentation vs. Sprinklers	Law allows compartmentation as alternative to sprinklers. Equivalent performance doubtful.
Compartment – Size	Allowable compartment size too large for manual fire suppression.
Compartmentation – Effectiveness	Fire spread vertically due to auto-exposure and through floor joints.
Smoke Detector Performance	Provided insufficient warning to prevent major fire involvement.
Fire Load	Heavy fire load in area of origin. Office cubicles with desktop computers and other equipment.
Ceiling Plenum	Heavy fire load with telephone and electrical cables. Fire may have originated above ceiling.
Defensive Strategy	Elevated master streams controlled fire. Would have been ineffective on higher floor levels.
Smoke Spread	Entire building filled with smoke.
Occupants	Unable to account for late night occupants, entire complex had to be searched.
Elevator Operations	Firefighter operating on manual control trapped, had to be rescued.
Stairway Pressurization	Ineffective against major fire combined with stack effect.
Structure	Columns and major beams undamaged. Light beams and floor deck warped and twisted.
Asbestos	Asbestos contamination complicated overhaul and investigation. All personnel and equipment had to be decontaminated.

LOCATION

The Banker's Trust Building actually consists of two major office towers, one 42 stories and the other 30 stories, linked by a 17-story connecting section to form an H-shape. (See Appendix A for site plan and floor plan.) This complex is surrounded by other highrise buildings in the middle of one of the most densely developed areas in the world. It is a major financial hub in National and international commerce, occupied by thousands of workers during work days. In addition, tens of thousands of New Yorkers and visitors pass by or under the complex every day; trains entering Grand Central Station, two blocks south, pass directly underneath.

The area is protected by some of the most experienced highrise firefighters in the world and one of the largest and most capable fire departments. Working fires in highrise buildings are not unusual in New York City, particularly midtown Manhattan. This fire, however, was significantly larger and more destructive than most and demonstrates the challenge that a working fire in a building of this size can create for a fire department; even a fire department with vast resources to engage in manual fire suppression.

SIMILAR INCIDENTS

This fire has remarkable similarities to two previous major highrise incidents, the One Meridian Plaza (Philadelphia) and First Interstate Bank (Los Angeles) buildings, and could have grown to similar magnitude under slightly different circumstances. All three of these fires occurred during the evening hours in office areas of highrise buildings constructed between 1960 and the early 1970's.[1] In all three cases the fire was discovered as a result of a smoke detector activation, yet there was major fire involvement prior to the arrival of the fire department, beyond the ability of a normal initial attack force to control.

The fire growth characteristics of the three cases all appear to be similar. The First Interstate Bank Building case is remarkably similar to this one, with respect to the type of space where the fires originated and the speed with which they reached flashover after activating smoke detectors. This evidence brings into question the effectiveness of smoke detection to provide early enough warning to protect highrise office buildings, particularly considering the time it takes for firefighters to arrive at the fire floor.

In each case security guards responded to the fire floor by elevator and were confronted with a significant working fire. The guard in Los Angeles was trapped in the elevator lobby and died, while the guard in Philadelphia had to be rescued. In this case the guard was able to leave the fire floor and advise other building staff that there was a serious fire in progress. This indicates that in all three cases fire growth was so rapid that occupants were quickly endangered.

Additional similarities include upward extension via auto-exposure (flames emanating from a lower floor breaking windows and entering the floor above) and through openings for electrical wiring. In two of the three cases there was damage to horizontal steel structure elements, but the vertical columns and major connections were not compromised. There is no evidence that massive structural collapse was imminent in any of the three fires.

The similarities provide an opportunity to compare the effectiveness of different code approaches to managing highrise fire risks. Each fire occurred in a major city that was able to send approximately 400 personnel to engage in or support interior manual fire suppression, and in each case the interior firefighting efforts proved to be ineffective. Very few cities in this country have the ability to assemble a firefighting force of this magnitude.

All three cities now require automatic sprinklers to be installed in new highrise buildings. In Los Angeles and Philadelphia the fires were the catalysts for adopting retroactive requirements to install automatic sprinklers in all existing highrise office buildings. In New York City Local Law 5 was adopted in 1973, requiring either sprinklers or compartmentation to be provided in existing structures. The compartmentation option had been selected and employed in this particular building.

FDNY HIGHRISE PROCEDURES

The New York City Fire Department has detailed policies and procedures for actions to be taken at highrise fires. Due to the number of incidents in highrise office buildings in the city (estimated at 300 actual fires per year), these procedures are well practiced and have been reinforced through experience. One of the standard operating procedures is that the first arriving ladder company is assigned to locate the fire while the remaining companies wait in the lobby for additional instructions.

---

[1] Highrise construction systems changed significantly in the 1960s and 1970s. Prior to this period, most highrise buildings were built with relatively heavy construction, providing a high mass to volume ratio which tended to provide natural compartmentation, heat absorption and insulation qualities. The newer buildings have much less mass – they utilize lighter weight steel or concrete structural members, curtain wall construction, more windows and thinner floor assemblies. All of these characteristics make the newer buildings inherently less fire resistive than their predecessors.

Another FDNY policy is to use no handlines smaller than 2-1/2-inch when working from stand-pipes. Standpipe outlets on the lower floors may have pressure limiting valves or restricted orifices to limit discharge pressures, but they are arranged so that the fire department is able to remove the restrictions when their hoses are connected to the outlets. This eliminates the problems that were encountered in the Los Angeles and Philadelphia fires with pressure control valves that restricted flows or failed to regulate discharge pressures.

When a working fire is encountered in a highrise building, a signal 10-76 is transmitted over the radio, upgrading the response to four engine companies, four ladder companies, one rescue company, four battalion chiefs, and a division chief. The Field Communications Unit, the Highrise Unit (a van carry extra equipment for use in highrise fires, staffed by Engine 3), a Command Post Company (one of several specially trained engine companies), and the Mask Service Unit (with additional air cylinders) are also dispatched. The full response to a 10-76 incorporates a total of approximately 73 personnel.

BUILDING DESCRIPTION AND SYSTEMS

The Banker's Trust Building is located at 280 Park Avenue,[2] between 48th and 49th Streets, in the heart of midtown Manhattan. The H-shaped building consists of two major office towers, one 42 stories and the other 30 stories which are linked by a 17-story connecting section. With the different heights of the towers and the connecting section and variations in the floor areas of the two towers, there are several different floor configurations at different levels.

The buildings were built in the early 1960s and are typical of highrise construction of that era. The exterior curtain walls are non-load bearing and include large windows separated by metal spandrel panels. The columns, girders, and beams are structural steel, protected in most areas by sprayed-on asbestos fire protective insulation. The floors are constructed of sheet steel decking covered with lightweight concrete ("Q-deck"). The central core is constructed primarily of steel encased in concrete with cement and gypsum block construction enclosing vertical shafts.

There are five enclosed stairways in the complex, labeled A through E. The west tower is served by stairways, C, D, and E. Stairway D, which was closest to the fire, is the smokeproof tower with vestibules vented to an enclosed smoke shaft located between the stairway and the occupied floor areas. All of the stairways are pressurized by fans that activate when a fire alarm is initiated.

Water is delivered via six-inch standpipe risers located in each stair tower, supplied by separate water supplies and fire pumps for each tower. There are also cabinets with 1-1/2-inch hose for occupant use in each tower on each floor.

Each tower contains both low-rise and highrise elevator banks. All of the floors involved in the fire are served by the low-rise elevators. The highrise elevators are located immediately to the west of the low-rise elevators and are in blind shafts that run through the fire area without openings. There are eight low-rise and eight highrise elevators, grouped in sets of four per common shaftway. There is also a single freight elevator which serves all floors, located to the west of the highrise elevators.

A corridor links the elevator lobby in the east tower with the west tower corridor system. There are automatic closing doors at each end of the corridor, forming part of the one hour compartmentation system.

---

[2]This location is directly across Park Avenue from the Westvaco Building, the scene of a major highrise fire in 1980.

There is a Fire Control Center in each tower. The west tower control center is located on the second floor, which serves as the main lobby floor for that tower. The east tower Fire Control Center is at the rear of the main entrance lobby on the ground floor, behind a bank of escalators. The normal path of travel for people entering the building is through the entrance lobby of the east tower and up the escalators to the second floor, which serves as the main elevator lobby level for both towers.

The presence of two Fire Control Centers caused a problem during the early stages of the fire, as the battalion chief initially assigned to this position did not realize that there were two. He stopped at the Fire Control Center that he encountered in the main lobby and did not realize for some time that the fire area, in the west tower, was controlled from a different location.

All fire protection systems for the two towers are maintained separately by design, to reduce the risk of total failure due to a single problem. The alarm systems in the two buildings are monitored by different central station alarm companies. (Both systems have additional annunciators at the single guard station in the main lobby near the Park Avenue entrance.) The buildings also have separate fire pumps and auxiliary generators.

THE FIRE

The fire at the Banker's Trust Building occurred on a Sunday night, January 31, 1993. At the time the building was occupied by a small force of security and maintenance employees, in addition to a few workers in their offices and personnel operating some of the 24-hour computer systems in the building. The point of origin was determined to be in the ceiling plenum above an open office area on the sixth floor and is believed to have been a result of overheated electrical wiring[3] igniting combustible insulation and other materials.

The area in question is an office used by the bank's international monetary trading brokers and included a large number of cubicle offices assembled of low height partitions and system furniture. Each workspace included some combination of desktop computers, monitors, printers, fax machines, telephones and other equipment, along with a generous supply of paper. The fuel loading was relatively high, there were numerous electrical devices, and the area was open, allowing for unrestricted rapid fire growth over a sizeable area.[4] When the fire began to involve the contents of the tenant space there was rapid fire growth.

The fire was detected at approximately 2243 hours by an ionization smoke detector in the immediate area. The smoke detection system sounded a local alarm in the area and on the floor above and at the building security station on the ground floor. At the same time the alarm was automatically transmitted to the central station alarm service that monitored the west tower. (A different company monitored the alarms in the east tower to provide additional redundancy.)

---

[3]The precise determination of the point and cause of the fire's origin was very difficult due to the extent of destruction and the fact that the area was contaminated by asbestos. The reported cause is listed as "probable" in the official report of the FDNY Fire Marshals who investigated the fire.

[4]The characteristics of the area of origin are remarkably similar to the area of origin of the First Interstate Bank Building fire in Los Angeles in 1988. See U.S. Fire Administration, Technical Report Series, Interstate Bank Building Fire, Los Angeles, California, Technical Report 22.

The central station alarm monitoring service called the Manhattan Communications Center of the New York Fire Department and advised them of a Class E smoke detection alarm on the sixth floor, west wing of the Bankers Trust Building at 230 Park Avenue. The call to the fire department was logged at 2247 hours. (See Appendix B for the Time Log of this incident.) The alarm service then called the building security station to advise that the fire department had been informed of the alarm and was en route. Both of these calls all appear to have been made within a period of 60 to 90 seconds, between 2245 and 2247 hours, although there may have been a discrepancy of as much as two minutes between the clocks at the different locations.

The security desk radioed the west tower security guard, who responded to the sixth floor by elevator to check on the alarm. The security desk also radioed the Deputy Building Safety Director, who was on duty in the east tower, and advised him of the alarm and asked him to respond to the lobby. The west tower security guard took one of the low-rise passenger elevators to the sixth floor and exited to find light smoke in the lobby. He noted heavier smoke in the corridor to the south of the lobby and then opened the door from the corridor to the office area. In the office area he encountered heavier smoke, with a layer of thick smoke moving along the ceiling from his right to his left (west to east). The guard was unable to see the fire, which he surmised was around the corner, somewhere near the west wall. The heavy smoke forced him to retreat, and he returned to the sixth floor elevator lobby where he advised the security desk by radio that there was a serious fire on the sixth floor.

INITIAL RESPONSE

The FDNY Communications Center dispatched Ladder 2, which was on the air returning from another alarm, and notified Battalion 8 of the call at 2248 hours. Ladder 2 arrived two minutes later at the front of the building on the Park Avenue side. There was no exterior evidence of a fire at that time. Ladder 2's crew was met at the front door by the Deputy Building Safety Director who reported that a security guard had already checked the sixth floor and there was an actual fire. At 2252 hours, Ladder 2 transmitted a 10-75 – a request for a full box alarm response of three engine companies and two ladder companies – and proceeded to the area where the fire was reported.

The crew of Ladder 2 used Stairway D to reach the sixth floor to determine the extent of the fire and the best attack route for the engine companies. This stairway is designed as a smokeproof tower with a vestibule between the stairway and the occupied area on each floor. The vestibule is vented to a smoke-shaft, which is designed to intercept smoke and keep it from reaching the stair enclosure. All three stairways in the west tower were also pressurized by fans that were activated by the alarm system.

The crew encountered moderate smoke while ascending the stairway and heavier smoke at the sixth floor level, coming from around the stairway door. Upon opening the door they encountered heavy smoke and attempted to enter by crawling low, using self-contained breathing apparatus (SCBA). The entry team penetrated approximately 15 feet down the corridor before being forced to retreat to the stair tower by heat and zero visibility. Ladder 2 then transmitted the 10-76, indicating a working fire in a highrise building.

FIRST ALARM ACTIONS

The 10-76 upgraded the assignment to a total of four engine companies, four ladder companies, a rescue company, four battalion chiefs, and a division chief. This was requested at approximately 2255 hours and the additional units were dispatched at 2257 hours. The 10-76 also brought the Field Communications Unit, a Mask Service Unit for additional air cylinders, Engine 3, which brings the High Rise Unit (a van carrying extra equipment for use in highrise buildings), and Engine 5, which is one of the companies specially trained to provide support for command post operations at highrise fires. The full assignment at this point in time included a total of 73 personnel, under the command of Deputy Chief Steven DeRosa of the 3rd Division.

As the 10-76 was being transmitted, Battalion 8, Chief Dawes, was arriving on the Park Avenue side of the building. There was no visible evidence of fire from that location and the battalion chief entered the lobby to establish a command post at the Building Fire Control Center, making contact with Ladder 2 on a tactical radio channel. He assigned Engine 8 to meet Ladder 2 at the sixth floor level in Stairway D to attack the fire.

Engine 65, the second due engine company on the first alarm assignment, responded from the west along 48th Street. As they approached the building they could see flames breaking through a row of windows on the sixth floor, moving from west to east. This observation was reported to Chief Dawes who transmitted a second alarm at 2258 hours.

Chief DeRosa, who arrived two minutes behind Engine 65 on the 48th Street side, noted that the fire was within reach of elevated stream appliances and special called Ladder 14, a 100 foot tower ladder equipped with a 2,000 gpm pump and twin master stream appliances. At this time flames were breaking out windows on the sixth floor across the full width of the building face on the south side and turning the corner on the east side.

When Engine 8 arrived at the sixth floor a 2-1/2-inch handline was connected to the standpipe in Stairway D and the door was reopened. Flames were encountered as soon as the door was opened and the line could not be advanced a few feet into the fire floor. A second 2-1/2-inch line was stretched by Engine 65 from Stairway E onto the fire floor. Only moderate smoke was encountered at the doorway onto the fire floor, but when the door was opened the crew encountered heavy heat and smoke. This doorway provided access to the opposite side of the building core and the line was stretched around to a narrow corridor that opened directly to the fire area.

Following the FDNY highrise procedures, two engine companies were assigned to each attack line. Engine 54 was assigned to work with Engine 8 and Engine 21 was assigned to work with Engine 65, but even with eight personnel rotating on each line they could not make progress into the fire area. Several of the personnel operating these lines received minor burns and all were exposed to punishing heat conditions. Engine 65's crew took a severe beating from the heat and smoke as they attempted to advance their line down the narrow corridor, but were unable to reach a point where they could hit the main body of fire.

Reports from the interior and exterior quickly presented a grim picture to Chief DeRosa who had assumed command of the incident. The fire was starting to break seventh floor windows and the threat of upward extension by auto-exposure was imminent. The interior attack crews were unable to advance far enough on the fire floor to make an offensive interior attack. In addition, Rescue 1, assigned to search the area immediately above the fire floor, reported that flames were coming up through cracks in the floor where the steel and concrete slab had split due to thermal expansion.

While the initial attack was being made, a full scale highrise fire command structure was established, with lobby control at the ground floor and a staging area on the fourth floor. Six battalion chiefs were assigned to operational positions, logistics functions, and planning/command support functions in accordance with the Incident Command System, while Division 3 assumed the role of Incident Commander at the building control station in the main lobby. Overall command of the incident was subsequently transferred to Assistant Chief Hughes who is normally in charge of the Division of Fire Prevention and was on duty as the designated Departmental Command Officer. Chief DeRosa took charge of Operations.

MULTIPLE ALARMS

The second alarm brought four additional engine companies, two more ladder companies, a second rescue company, another battalion chief and a second division chief, and Engine Company 9 with Satellite 1, a large volume water supply company. The on-duty battalion chief from the Division of Safety and the Citywide Command Chief also responded. The second alarm brought an additional complement of 51 personnel to the scene.

At 2321 hours a third alarm was transmitted by Chief DeRosa. Four more engine companies, another tower ladder, two additional battalion chiefs, the Mobile Medical Unit, and two more large volume water supply companies responded on the third alarm, with another 42 personnel.

DEFENSIVE STRATEGY

The concern for upward extension mandated a switch from offensive to defensive strategy. The interior companies were instructed to back out to the stairs in a holding action, in anticipation of an exterior attack while tower ladders were set up on the south side of the building for an exterior attack. Tower Ladder 7, a 75-foot platform, was already in position, waiting for word that the interior crews were in safe positions, while tower Ladder 14 was just setting up. The 75-foot tower ladder could barely reach the sixth floor level, while the 100-foot platform could reach the eighth or ninth floor. Satellite 1, a hose wagon with a giant monitor nozzle (originally obtained with the Super Pumper System in 1963), also set up for master stream operations.

When all interior crews reported that they had pulled back down to the fifth floor and accounted for all crew members, the order was given to attack the fire from the outside. The elevated streams were then used to deluge the sixth and seventh floors, first to stop the threat of exterior auto-exposure and then to knock down the flames on both the sixth and seventh floors.

Two additional tower ladders were special called at 2330 hours to provide more elevated stream capability. These tower ladders were to be set up on the 49th Street side in case the fire extended to the north half of the west tower. A fourth alarm was transmitted seven minutes later, bringing 26 additional personnel to relieve the initial attack companies, many of whom were in need of medical evaluation for exhaustion and minor injuries.

SEARCH AND RESCUE

Since this fire occurred late on a Sunday evening, the initial information suggested that there would be few occupants in the building. The first priority for interior search was the floor areas adjacent to and immediately above the fire area; these areas were searched for occupants by first and second alarm companies. As the search was being completed in these areas, a call was received at the Manhattan Fire Communications Center from an occupant who reported that he was trapped by heavy smoke above the fire. A rescue team was dispatched to rescue this occupant, while an assessment of the potential for additional occupants was conducted.

The firefighter assigned to operate the elevator that was used by the rescue team became trapped when the elevator he was operating stopped on the seventh floor. At that time the fire had extended to the seventh floor and the area was heavily charged with smoke and hot gasses. He transmitted a "mayday" over his portable radio to inform the Incident Commander that he was trapped. Rescue Company 1, which happened to be on the seventh floor at the time, heard the mayday and rescued the trapped firefighter within a few minutes.

A check of the building log indicated that several occupants had signed in during the day and had not signed out on the log sheet at the security desk. A decision was made that a full scale search of the entire complex would have to be conducted. Crews already on the scene or responding on the fourth alarm were initially assigned to conduct this floor by floor, room by room search.

As the magnitude of this task became clear, the fifth alarm was requested for additional crews at 0024 hours. A dozen additional fresh crews were requested more than an hour later to provide further assistance for the search. Over a period of three hours the entire floor area of both towers was searched and ventilated. No additional occupants were found.

At 0430 hours, six additional relief companies were requested to take over salvage and overhaul as the companies on the scene were decommitting. The total response equaled eight alarms and over 400 fire department personnel. Approximately 35 fire department members were injured, primarily with minor burns and other injuries, as well as extreme fatigue and exhaustion. A few of the firefighters were admitted to hospitals for observation, but no major injuries were reported.

STRUCTURAL CONCERNS

As the initial entry was made to overhaul the fire area, deformed and sagging structural steel beams were noted and all personnel were withdrawn from the area. Rescue Company 3, which specializes in collapse rescue operations, was special called to the scene at 0310 hours to provide temporary shoring. Overhaul was resumed after temporary supports had been installed in the weakened area.

Most of the structural damage was confined to the underside of the seventh floor assembly. There was minor warping and sagging of the steel beams and larger girders, as well as sagging of the steel Q-deck between the beams. One of the smaller beams was detached from the supporting girder at one end of its span, apparently from the force of a master stream hitting it while it was heated. The deformation of the deck was most pronounced along the joints where fire had been observed extending through to the seventh floor; a variation in the floor level of three to six inches remained after the fire.

Close examination after the fire was completely extinguished showed there was no damage to the columns or girders and the structural deformation of the floor assembly was relatively minor. Several floor sections and a few beams would need to be removed and replaced to repair the structure. This was similar in nature to the damage that was observed in the highrise office building fires in Los Angeles and Philadelphia, although the extent of the damage in Philadelphia was much greater.

DAMAGE ASSESSMENT

The direct fire damage on the sixth floor included destruction of all contents on the south half of the west tower and heavy heat and smoke damage on the north half of the same tower. A smaller area directly above was involved on the seventh floor and there was minor fire extension into the eighth floor. Most of the remainder of the sixth floor of both towers was heavily damaged by smoke, although the area separation doors kept the flames out of the east tower and limited the heat damage.

The master streams also punched a hole in the block wall enclosing the elevator shafts, directly opposite the windows, causing the blocks to fall on top of the elevator cars that were parked at the lobby level. Water runoff damaged most of the lower floors of the west tower. Electrical and telephone systems in the west tower were disrupted and needed major repairs.

All of the floors above the fire, in both towers, were smoke filled and also considered to be potentially contaminated by asbestos. Contractors were called in to initiate a massive clean up operation, trying to limit the damage to computers and other electronic equipment from the soot that settled on every surface. The direct damage estimate was well in excess of $10 million and parts of the building were shut down for several weeks, adding to the economic loss.

ASBESTOS CONTAMINATION

An additional problem was encountered during the overhaul stage, when it was determined that the fire area was contaminated by asbestos. The structural steel above the ceiling had been protected by sprayed on asbestos, most of which was washed off by the water streams and had fallen on top of the fire debris. Additional asbestos may have been carried by convective air currents throughout the building. All of the firefighters who worked inside the building and all of their protective clothing, breathing apparatus, tools, and equipment were presumed to be contaminated. The entire west tower was suspected of being contaminated, due to the amount of asbestos that was loosened from the steel structure.

The FDNY's decontamination unit, a 40-foot trailer operated by Ladder Company 15, was special called to the scene at 0602 hours to process all of the personnel and then all of the equipment through asbestos decontamination. This process took several hours.

The asbestos problem continued through the investigation stage. Investigators could not enter the fire area to determine the cause and origin of the fire until the asbestos problem had been addressed. The building owners brought in an asbestos removal and decontamination contractor to remove the free asbestos from the fire area and from the rest of the building; however, it was three days before investigators were able to enter the area wearing respirators and protective ensembles.

HALON DISCHARGE

It was also discovered during the overhaul process that at least two Halon 1301 fire extinguishing systems had discharged in the building, one in a room on the sixth floor that was not involved in the fire and the other on a higher floor. The Halon systems were designed to protect computer rooms and were activated by ionization smoke detectors in the protected spaces. There was sufficient smoke penetration into these spaces to cause the systems to discharge. This did not cause any problems in the buildings and because of their locations; the systems had no effect on the fire. An unnecessary discharge of Halon 1301 to the atmosphere, however, is undesirable because of its damaging effect on the ionosphere.

ANALYSIS

This section of the report identifies and analyzes the significant issues that can be drawn from this incident.

Local Law 5

Local Law 5 was adopted in 1973, following a series of major highrise fires in office buildings in New York City. After adoption, its enforcement was delayed for five years while building owners challenged the authority of the City of New York to enact retroactive requirements for existing buildings. After the law was upheld, an additional law was passed to establish a timetable for implementation of the requirements of Local Law 5. The compliance program required existing buildings to be in full compliance by the end of 1983, if automatic sprinklers were to be installed, and by the end of 1988 if the compartmentation option was selected. The buildings were required to complete work in phases to ensure that progressive efforts would be initiated in the early years of the time allowed for full compliance.

Local Law 16, adopted in 1984, requires automatic sprinklers to be installed in all areas of new buildings over 75 feet in height. Local Law 5 applies only to buildings built before this requirement came into effect.

The most stringent requirements of the Local Law 5 are directed toward buildings more than 100 feet in height that have air handling systems serving more than one floor. These buildings were required to install automatic sprinkler systems on all floors or to divide floor areas into fire resistive compartments. Approximately 275 buildings came under the requirement to select one option or the other. Approximately 100 buildings were already protected by sprinklers at the time the law came into effect. Extensions to the compliance deadline were requested by 33 buildings and most of these buildings are still in some stage of appeal or partial compliance.

Where the compartmentation option was used, individual floor areas were required to be divided into compartments not exceeding 7,500 square feet, using one-hour rated wall assemblies. An optional configuration allowed compartments of up to 15,000 square feet, divided by two-hour rated fire resistive construction, if smoke detection systems are also installed within the compartmented areas. The areas specified in each case are net occupied floor areas, exclusive of exit corridors, stairways, elevator shafts, and other core areas. Where the compartmentation option is selected, the law also requires a smoke control system, which may be a smoke shaft, pressurized stairways, or an equivalent engineered system.

Elevator recall and emergency operation systems, elevator lobby and stairway labeling systems, and Class E alarm systems are also required. The elevator control system requirements include smoke detectors in the elevator lobbies on each floor to recall all cars to the ground level (or the lowest practical level) and manual controls for firefighters to operate the recalled elevators. Class E alarm systems include public address system capabilities on each floor and two way communications to the building's Fire Control Center, as well as annunciation of the required smoke detection systems at the Fire Control Center. There are several variations of Class E alarm systems for different situations.

Local Law 5 also required each building to appoint a Building Fire Safety Director to manage the fire safety program for the building. A Building Safety Director or a Deputy Director must be on duty at the premises when the building is occupied and is required to respond to the Fire Control Center to meet the fire department whenever an alarm is sounded. Both Building Safety Directors and Deputy Directors must be trained to meet fire department requirements and are tested and specifically certified for the particular building by the High Rise Section of the Division of Fire Prevention. There are approximately 5,000 certified Directors and Deputy Directors in New York City.

When a building is occupied by fewer than 100 people the responsible individual on duty may be a Building Evacuation Supervisor. Building Evacuation Supervisors are normally trained in-house by the certified Director and usually have other duties to perform.

The Fire Control Center is usually located on the ground floor and must contain the fire alarm annunciator and control panels for the public address and internal communications systems, as well as floor plans, evacuation plans, and information on fire control systems that are installed in the building. The Building Safety Director (or the designated individual on duty) is expected to stay at the Fire Control Center to work with the fire department, providing information on the protective systems in the building and helping to manage evacuations.

Effectiveness of Fire Protection Concept of Local Law 5

At the time of its adoption, Local Law 5 was extremely controversial. The New York City Fire Department, headed by Commissioner and Chief of Department John T. O'Hagen, held the position that all new and existing highrise buildings should be protected throughout by automatic sprinklers. The compartmentation options were the product of a hard fought compromise with the building owners who resisted any retroactive requirements. Even after adoption by the City Council the law was challenged all the way to the State Supreme Court before it could be enforced.

The level of protection provided by compartmentation assumes that a fire can be successfully contained to the defined area by a combination of passive fire resistive construction and the manual fire suppression efforts of firefighters. With the largest on duty force of firefighters in the United States, the New York City Fire Department has the ability to commit an unparalleled manual fire suppression capability on a structure fire. It has been debated for years whether or not this force would be adequate to control a fire of the dimensions allowed by Local Law and whether or not the compartmentation systems would be effective.

A statistical study, that was specifically directed toward this question, concluded that sprinklers were significantly more effective than compartmentation as a means of controlling fires in highrise office buildings.[5] The study looked at 1,530 reported fires in New York City, during the period from 1981 through 1985 and concluded that the fires in compartmented buildings damaged a larger proportion of the structure and contents, lasted longer, and required the efforts of more firefighters, more engine and ladder companies, and more hoselines to bring them under control.

The conclusions of the officers who fought this fire was that had the fire occurred above the reach of the exterior streams, the fire could well have continued uncontrolled and involved the floors above the fire. That the fire was successfully fought was, in part, due to the fact that it occurred on one of the lower floors. Interior attack was incapable of controlling the fire on the sixth floor and could not prevent extension to the seventh and eighth floors. The exterior elevated streams that did control the fire were effective at the sixth and seventh floor levels and could have been used up to the ninth or tenth floor, but on higher floors they would not have been capable of gaining control of the fire.

Above the reach of aerial apparatus, extension by auto exposure could possibly have been limited by streams projected upward from tower ladders and by the satellite monitors, possibly up to above the fifteenth floor, but there would be reduced penetration of the streams to knock down the interior fire

---

[5] *An Effectiveness Comparison of Sprinklers and Compartmentation for High Rise Office Building Fire Protection as Defined by Local Law 5 (1973) for the Years 1981-1985*, Charles Jennings, John Jay College of Criminal Justice of the City University of New York.

on each floor. Beyond that height the opportunities for exterior attack become limited, to only such methods as projecting streams from the windows or rooftops of nearby buildings.

While there is a chance that auto exposure could be limited in this manner, under the most optimistic circumstances, it is very doubtful that interior extension could be limited at the same time. This fire was at the point of extension into the seventh and eighth floors, through existing penetrations in the floor assemblies and the expansion joints, and the structural conditions required the withdrawal of firefighters from the weakened areas, when it was controlled by the exterior streams. The evidence strongly indicates that this fire would have extended vertically if the exterior attack had not been effective.

This conclusion again calls into question the basic premise of Local Law 5. It appears to confirm that there is a real possibility of a much larger and potentially uncontrollable highrise fire in a building that relies on compartmentation as the basis of fire protection. The assumptions that a combination of passive structural fire protection and manual fire suppression will successfully confine a fire to 7,500 or 15,000 square feet appear to be overly optimistic, judging from this experience. With lightweight construction and heavy fuel loads, the compartments appear incapable of containing a fire and the fire volume exceeds the capabilities of interior attack forces. This reinforces the opinion that automatic sprinklers are a superior and more reliable form of fire protection for highrise buildings.

Automatic sprinklers have been accepted as the standard of protection for new and existing highrise office buildings in most parts of the United States at the present time. The analysis of this fire appears to support the conclusion that compartmentation does not offer equivalent protection in highrise buildings, particularly those of modern lightweight construction.

Automatic Detection versus Automatic Sprinklers

A second companion debate has centered on the question of the relative effectiveness of smoke detection systems versus automatic sprinklers in protecting commercial properties and particularly highrise buildings. This incident suggests that smoke detection systems may not offer sufficiently early warning to allow firefighters to respond and control fires before they reach an uncontrollable magnitude. The circumstances of this incident suggest that this condition occurs when a flashover condition occurs in a space that is larger than a fire department can reasonably attack and control with handlines.

The tenant space on the sixth floor of the west tower was equipped with a smoke detection system that was in addition to the fire protection systems required by the Local Law 5 or any other codes that applied to the buildings. A total of 62 ionization smoke detectors were installed in this space, with an average coverage area of approximately 140 square feet per detector. This could be described as a fairly standard distribution for this type of detector.

In addition to the detectors protecting the tenant space, there were three detectors located above the suspended ceiling, in the plenum area, and detectors in the elevator lobbies. The detectors in the plenum space were designed to shut down the return air systems if smoke was detected in the airflow from the sixth floor into the vertical return air shafts.

The air handling system in the building was a fairly common type for this type of building, designed to deliver cooled or heated air to the tenant spaces through ducts and to allow the return air to flow through the plenum space back to the return air shafts. Cooling and heating were provided through large systems located on mechanical equipment floors. With the air handling system in operation, smoke originating from the sixth floor area would be drawn past these detectors, which were designed to shut down the system. Two of these detectors were sampling tube type detectors located in the air stream, while the third was a regular area coverage detector located in the air path to a return air shaft.

Detectors located in return air flows have been shown to be relatively insensitive to smoke generated within the space from which the returning air is taken, since there are generally massive quantities of air passing over the detector, which highly dilutes the products of combustion, and the air is moving at high velocities, which makes it difficult to detect small quantities of smoke. These detectors are installed to shut down the recirculating air system if there is a large quantity of smoke entering the system – not to be a primary form of protection for the space.

At the time of the fire, the main air handling systems were shut down. Under this condition the only airflow into or out of the return air shafts would be the natural flow resulting from differentials in air density caused by temperature variations and stack effect. Due to the cold outside temperatures at the time of the fire, the expectation would be a moderate upward flow, drawing air into the shafts at the lower floors and out at the upper floors. Many other factors can influence airflows within plenum spaces and shafts at different times; however, it is impossible to be certain which may the air would actually flow and the velocity of the air under shutdown conditions.

The sequence of detector zone activations, taken from the master control printer, indicates that the first ionization detector, located in the tenant space, activated at 2243 hours. Additional detectors in the same space may have operated in rapid succession, but they would not be logged since the detector zone was already activated. Seven minutes later, at 2250 hours, the elevator lobby detector on the sixth floor activated, indicating that smoke was now spreading beyond the tenant space. One minute later the duct detectors on several floors began to activate and over the next seven minutes more than 20 additional detection zones activated, from the fifth floor to the roof level. By this time, massive quantities of smoke were moving upward through the structure, primarily through the return air shafts and possibly through the elevator shafts.

This sequence coincides with the observations of the security guard who observed only light smoke on the sixth floor when he reached the elevator lobby, responding to the initial alarm, but rapidly building smoke in the tenant space. By the time Ladder 2 was arriving at the front door, only seven minutes after the first detector had activated, the smoke was spreading rapidly and within the next two minutes the fire had reached a point that heavy smoke was spreading rapidly through the return air system. With the fans shut down, the combined forces of the heated smoke and stack effect would cause the smoke to flow into the return air shafts and quickly activate the detectors at the entrances to the shafts on other floors. This sequence can be seen from the following table which shows the sequence of detector zone activations.

Smoke Detector Activation Sequence

TIME	FLOOR	DEVICE
2243	6	Tenant Area System
2250	6	Elevator Lobby
2251	10	Duct Detector
2251	11	Duct Detector
2252	6	Duct Detector
2252	12	Duct Detector
2252	5	Tenant Area System
2253	16	Duct Detector
2253	7	Tenant Area System
2254	22	Tenant Area System
2254	18	Duct Detector
2255	16	Tenant Area System
2255	26	Elevator Lobby
2255	23	Tenant Area System
2255	7	Telephone Room
2256	28	Elevator Lobby
2256	28	Tenant Area System
2256	29	Tenant Area System
2256	14	Tenant Area System
2256	9	Tenant Area System
2256	Roof	Stair Pressurization Air Intake

The sequence of events indicates that flashover was occurring in the tenant space within no more than 14 minutes after the initial detector activation and probably in 10 minutes or less. It is unreasonable to expect that the fire department will arrive in less than ten minutes to an alarm from a space of this nature, considering the time it takes to transmit an alarm to the fire department, for a company to respond to the building, and then for personnel to reach the reported area within the building. In a highrise building it may take additional time for companies to verify the location at the lobby before proceeding to the reported fire floor.

The most significant difference between smoke detection and automatic sprinklers is clearly evident when time is considered. While the detectors may activate more quickly and give an earlier alarm, they only warn the occupants and summon the fire department. Automatic sprinklers have the added capacity to control or extinguish a fire before the fire department arrives and have proven to be highly reliable and effective.

Response to Class E Alarms

There are approximately 1,200 highrise office buildings in New York City, with the great majority concentrated in Lower and Midtown Manhattan. Because of the requirement of Local Law 5 to install smoke detectors in all unsprinklered areas where the compartmentation exceeds 7,500 square feet, as well as in all air handling systems and elevator lobbies, these buildings contain tens of thousands of smoke detectors, all of which are connected to central station alarm monitoring services. Many tenant areas also have smoke detection systems installed, even where they are not required by codes. These smoke detection systems generate hundreds of alarms every month, the vast majority of which are found to require no fire department intervention.

The smoke detector alarms created such a high activity level for the fire companies in the primary highrise areas, to the extent that their availability for other incidents was sometimes compromised and their crews were becoming fatigued from making so many runs. After analyzing the pattern of smoke detector alarms, the New York City Fire Department created the category of "Class E response" and implemented a policy of sending only a single engine or ladder company to smoke detector alarms from office buildings, when no additional indication of a fire is received. The battalion chief is notified of the alarm and may or may not respond.

One of the factors influencing the adoption of the Class E response policy was that in those cases where an actual fire was found, the notification from the central station was almost always followed quickly by at least one supplementary telephone call to the fire department from the building, before or soon after the automatic alarm notification was received. The building safety directors are instructed to call the fire department immediately if an actual fire is discovered and to meet and assist the responding companies at the designated entrance.

Over a period of more than ten years the single company response policy has proven effective in reducing the number of unnecessary responses for companies in the highrise districts. It does not appear to have compromised effective response to those situations where intervention was required. In the analysis of this incident it appears that the response of a single ladder company on the smoke detector alarm did not significantly delay the attack on the fire. Ladder 2 arrived on the scene within three minutes and was met by the Deputy Building Safety Director. The company officer was notified that the security guard had reported actual smoke on the sixth floor and the full box alarm response was immediately requested, calling for three engines and another ladder.

Engine 8 was on the scene and the remainder of the first alarm units were arriving by the time Ladder 2 reached the sixth floor and reported heavy smoke. The minor delay factor for the arrival of the remainder of the first alarm companies does not appear to have been significant in the outcome of the incident.

Smoke Movement

Several factors influence the movement of smoke within a highrise building. Stack effect, which is caused by differentials in the density of air at different temperatures, is particularly significant when the exterior is much colder than the interior of a highrise building, as was the case at the time of this fire. Stack effect tends to draw air from the lower parts of a building, where there is inward leakage, to the upper levels where there is outward leakage. When the normal air handling systems are shut down, the drafts caused by stack effect would be expected to cause air and smoke to move upward in the return air shafts.

In the initial stages of a fire, the movement of smoke tends to follow the natural air currents, until the fire generates sufficient heat to expand the air volume and create its own convection currents. At that point the forces created by the fire become significant and the buoyant air mass flows outward and upward, creating its own pressures which interact with stack effect and other air currents. The hot smoky air mass will flow toward air shafts that are at a lower pressure and provide a path to the upper levels of the building. This explains the rapid activation of the smoke detectors at the return air shaft entrances on floors above the fire.

The huge volume of smoke generated by this fire quickly spread throughout the complex above the sixth floor. When the windows on the sixth floor were broken by the heat of the flames, the prevailing wind pushed fresh air into the fire area and helped to contain the smoke within the structure. When firefighters opened stairway doors to make entry to fight the fire, the smoke was able to enter these shafts and the flow of smoke and hot gasses toward the doorways made it more difficult for firefighters to advance hoselines into the fire area. This appeared to be a very significant factor in the case of Stairway D, since the smoke and heated gasses were drawn toward the smoke shaft; this put the attacking firefighters in the path of the heat and smoke flowing toward the smoke shaft and made it even more difficult for them to advance handlines into the fire area.

Stairway Pressurization

The stairway pressurization systems, which are designed to keep smoke out of the stairways, did not appear to be effective in this incident. All of the stairways were equipped with pressurization systems, which are activated by any fire alarm condition. Stairway pressurization is intended to counter the flow of smoke into the stair shafts by delivering pressurized outside air to the stair shafts, creating a higher pressure inside the shafts than in the adjacent floor areas. The design criteria call for a pressure differential to be maintained with a number of doors open, allowing the pressurization air to flow out.

It is very difficult to design a system that can balance the pressurization against multiple doors being open at the same time, and also account for stack effect and pressures resulting from the expanding air mass created by a hot fire. When windows are broken, as they were in this fire, additional pressure conditions may be created by the wind. There is also an indication that a smoke detector on the fresh air intake that supplied the air for stairway pressurization was activated and shut down one or more of the pressurization fans fairly early in the incident.

It appears that this type of pressurization is ineffective against a fire that is generating significant volumes of smoke and creating its own pressures, particularly when the door to the fire floor is opened and kept open for firefighting operations. Stairway pressurization should be somewhat more effective in keeping smoke out of an alternate stair shaft to facilitate occupant evacuation, if the door from that shaft to the fire floor is kept closed. In many cases it would be effective to designate one stair tower for attack and to anticipate that it will be contaminated by smoke and to keep others clear for occupant evacuation. The "smokeproof tower" is usually preferred for evacuation, since it is designed to keep smoke out of the stairshaft.

An alternate configuration for a pressurization system, designed to direct a flow of pressurized fresh air directly against the shaft side of each door opening, may be more effective than mass pressurization of the entire shaft. This type of system is much more complicated and more expensive, however, and requires additional space for a fresh air supply duct or shaft within or adjacent to each stair shaft.

The problems identified with stairway pressurization are much less significant when the building is equipped with automatic sprinklers, since the fire will not usually create smoke and heat conditions that challenge the effectiveness of the pressurization system. The smoke created by a fire in a sprinklered building is usually "cold and wet" (i.e., close to the ambient temperature and saturated by the water application) and does not expand in volume; therefore, convection currents and expansion pressures are much less challenging.

Occupants

Even with effective fire resistive compartmentation, the massive smoke spread throughout this complex resulted in a major property loss. The smoke could also have resulted in many more casualties if the building had been more heavily occupied at the time of the fire. During work hours there would normally be three to four thousand occupants in this complex above the sixth floor, all of whom would have been endangered by the fire and products of combustion.

Experience has shown that fires of this magnitude are much more likely to occur at night or on weekends, when discovery of the fire depends on automatic detection systems, but even at these times it is not unusual to find dozens of workers operating 24 hour computer systems, conducting transactions with brokers on distant continents, or conducting construction or maintenance activities in the building. The fire department must have a plan to locate and evacuate all occupants, whenever a fire occurs.

The log-in and log-out system that was in use at the Banker's Trust Complex did not accurately account for the occupants of the building during the late night hours. The problem was recognized when the call was received from a trapped tenant on an upper floor, and the uncertainty made it necessary to search the entire area of every floor, room by room, to check for occupants. This took several hours and the commitment of more than 20 companies.

Elevators

The use of elevators under fire conditions continues to be controversial, as demonstrated in this incident. In this incident elevators were used to transport personnel above the fire to search for reported occupants on the upper floors, but a firefighter operating an elevator became trapped and had to be rescued.

The use of elevators during fire conditions is still debated by fire departments; some have determined that they will never use elevators, while others consider them to be part of a standard attack plan. The New York City Fire Department routinely uses elevators and has standard operating procedures that define when, how, and by whom elevators are to be used. In this case an elevator that served the fire area was used, which violates one of the basic procedures.

One of the defined functions of the command post companies is to provide elevator operators. It was determined that the firefighter who was operating the elevator had responded on a designated command post company and should have been specifically trained to perform this function, but was only temporarily detailed to the company. The individual did not have an SCBA or a forcible entry tool in the elevator, both of which were required by standard operating procedures. Fortunately, he did have the required portable radio and was able to call for assistance, and a rescue company was working in the immediate area where he was located.

Other Problems Encountered

Several notable problems were encountered in this incident and were being considered as lessons learned by the New York City Fire Department.

Multiple Stairways – The decision on which stairs to use for fire attack and which to use for evacuation should be based on the specific location of the fire and the options that are available in each building. The twin tower complex included five major stairways and the smokeproof tower was selected as the primary fire attack point due to its proximity to the reported fire area. This placed the attack crews in the path of the smoke and heat that were drawn toward the exhaust shaft when the doors were opened, drawn by stack effect and pushed by the wind. The selection of a different stairway for the attack might have provided a less punishing approach, as well as maintaining the smokeproof tower free of smoke for the evacuation of occupants.

Multiple Fire Control Stations – The twin tower complex at the Banker's Trust Complex created a problem, as the two towers were considered as independent for all fire and safety systems. There were two Fire Control Stations; the one in the main lobby served the east tower, while the one for the west tower was on the second floor. The Battalion Chief who was initially assigned to the Fire Control Station found the more visible one and did not realize there was a different location to monitor and operate the systems in the fire area. For a short period of time there were two command officers at two different Fire Control Stations.

Floor Plans – The building owners are required to provide up-to-date floor plans at the Fire Control Station; however, it was difficult to determine the perimeters of the designated fire compartments on each floor from the plans that were provided (even after the fire). It was known that the floors were compartmented, but it was difficult to tell where the compartment lines were located from the plans.

Detailed Personnel – As noted in the discussion on elevators, some of the designated special function companies were operating with several detailed personnel and were limited in their ability to perform the functions that are normally assigned to them. This problem may be addressed by providing additional training for personnel who can be detailed to the designated companies, or by temporarily suspending their special qualifications when they have a shortage of trained personnel on duty.

Fuel Load in Ceiling Plenum – The space above the ceiling in the fire area and adjacent areas was used for electrical wiring and communications cables. The amount of communications wiring in this type of occupancy is often a problem. It appears that over the years the amount of communications wiring in the plenum has grown, as old runs have been replaced by new wiring, but the old cables were not removed. The fuel load provided by the insulation on the wires contributed to the fire origin and intensity. The cable runs also made it impossible to fully seal the openings in fire separation walls above the ceilings, allowing smoke to migrate from one compartment to another and contributing to the secondary damage.

LESSONS LEARNED

1. **Effectiveness of sprinklers versus compartmentation.**

   The most important lesson that should be derived from this incident is the strong indication that a compartmentation strategy does not provide an equivalent level of protection to automatic sprinklers for highrise structures. Even when smoke detection is installed to provide early warning, the real possibility exists that a major fire will occur in a large space that is heavily loaded with combustibles. In this case the smoke detectors did not provide sufficient warning for the fire department to respond and control the fire before it reached major proportions.

2. **Compartment size.**

The compartment size allowed by Local Law 5, up to 15,000 square feet per floor, is too large for manual fire suppression to be effective in many cases. If a compartment of this size becomes involved, or even a major portion of a large compartment, particularly modern buildings of "lightweight" construction, there is a significant risk that the fire may spread to higher floors or adjacent compartments. Once a fire spreads beyond the floor of origin, the challenge to manual fire suppression forces is extreme and vertical spread may be uncontrollable, if it occurs beyond the reach of elevated stream apparatus.

3. **Smoke spread is a major problem.**

This incident again demonstrates the manner in which smoke can spread rapidly throughout a highrise building, when an uncontrollable fire is in progress. The stack effect caused by temperature differentials and wind forces can add significantly to smoke spread problems and overwhelm the ability pressurization systems to keep smoke out of stairways. Uncontrolled penetrations of communications wiring and other combustibles in ceiling plenum spaces can provide an avenue for smoke, as well as fire, between compartments.

4. **A working highrise fire places tremendous resource requirements on fire departments.**

This fire once again demonstrates the massive numbers of firefighters who are needed to control a working highrise fire, in both suppression and support roles. The logistical functions are equally demanding to direct suppression activities and both are likely to result in numerous injuries, particularly due to fatigue.

5. **Elevator use during highrise fires requires special training and precautions.**

Even with elevator controls designed to be used by firefighters, an untrained firefighter found himself in a critically dangerous situation and had to be rescued.

6. **Command and control of highrise fires requires a high level of coordination and information.**

Well trained building personnel can be a major asset to the fire department in managing a building and its mechanical systems during a fire.

7. **Asbestos contamination can seriously complicate an already complicated operation.**

Fire departments need to be prepared to deal with asbestos contamination as an additional complication in major fire incidents.

APPENDIX A

Banker's Trust Complex

Site Plan, Sixth Floor Plan, and Fire Attack Plan

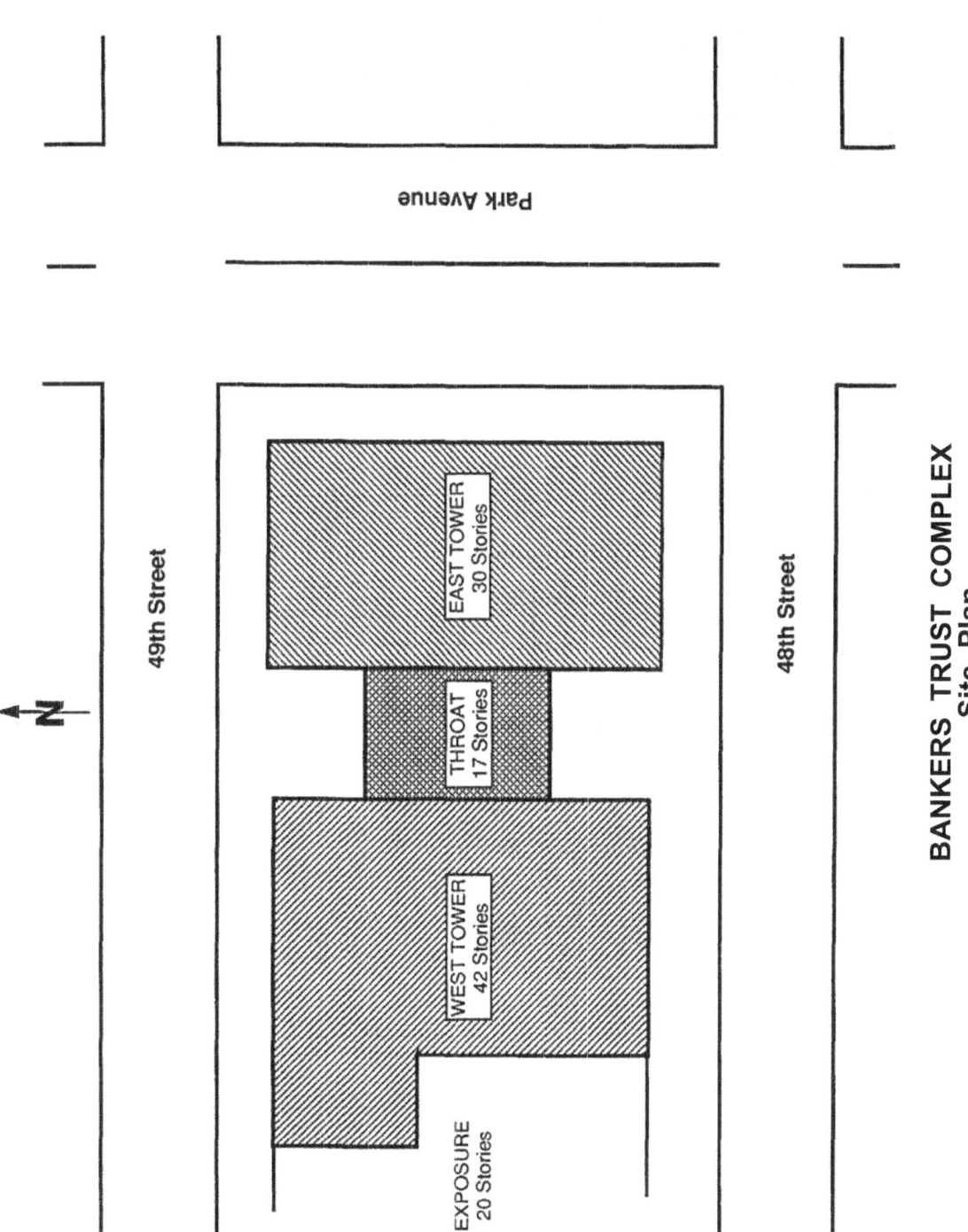

BANKERS TRUST COMPLEX
Site Plan

Appendix A (continued)

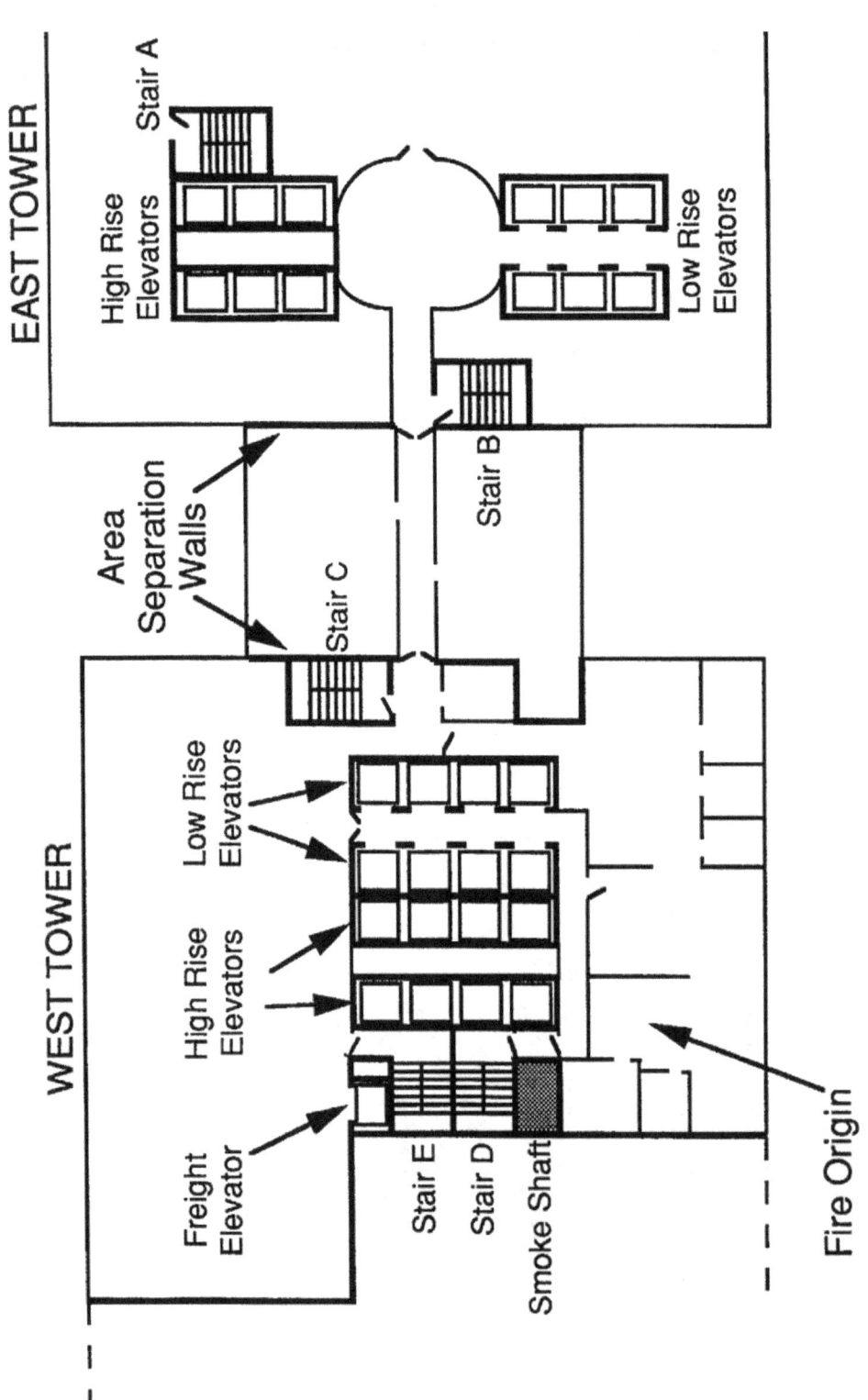

**BANKERS TRUST BUILDING
FLOOR PLAN - 6TH FLOOR**

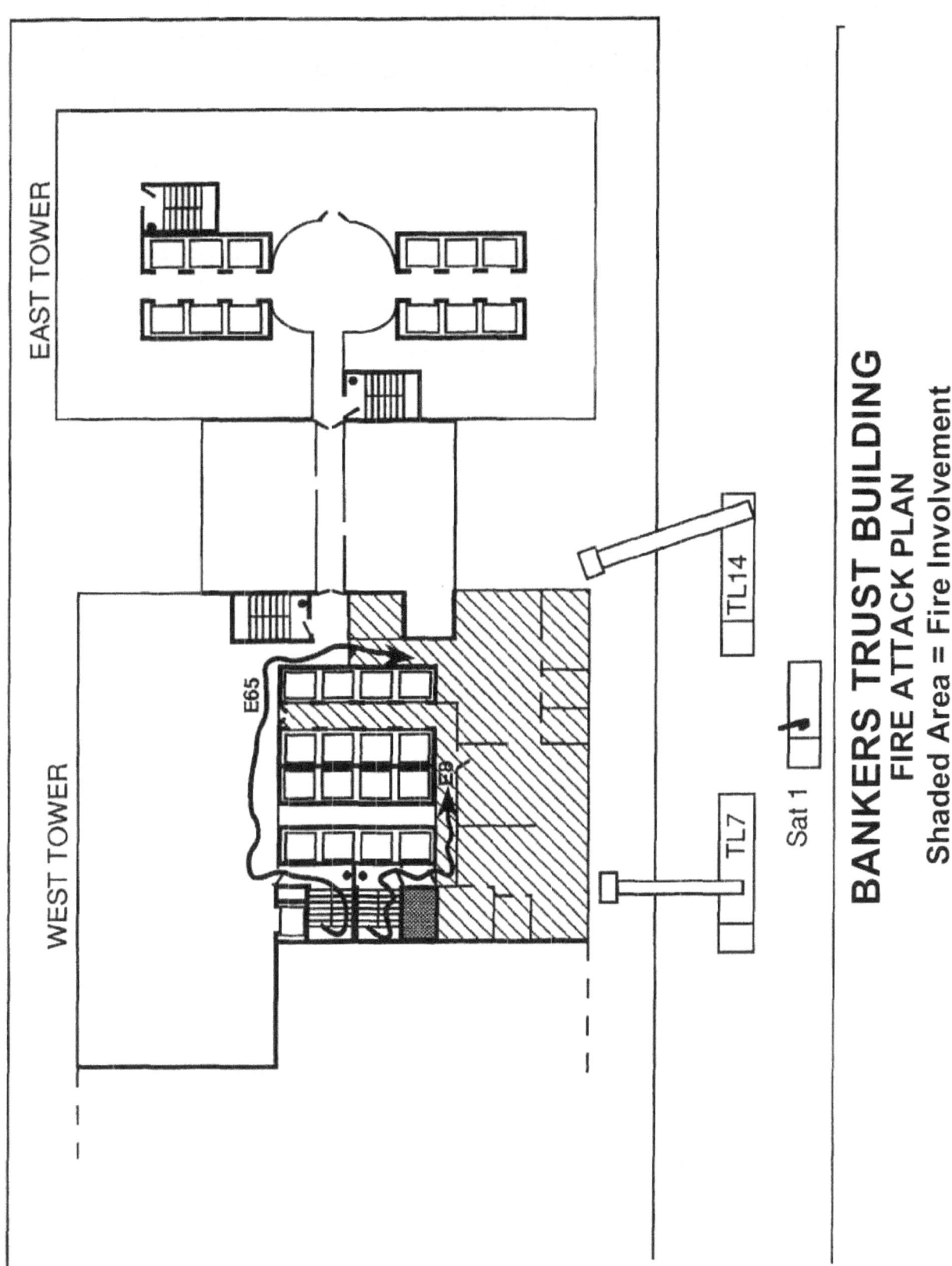

Appendix A (continued)

BANKERS TRUST BUILDING
FIRE ATTACK PLAN
Shaded Area = Fire Involvement

EAST TOWER

WEST TOWER

E65

TL14

TL7

Sat 1

APPENDIX B

Time Log of Banker's Trust Building Fire

January 31, 1993, New York City

TIME LOG – 280 Park Avenue

TIME	ACTIVITY
2247	Class E Alarm received from AFA for 280 Park Avenue, West Tower
2248	Dispatched L2 (on the air), BC8
2250	L2 10-84
2252	L2 Requested Full Assignment (10-75) Dispatched Box 827 48th Street & Park Avenue E8, E65, E54, L4
2255	BC8 10-84
2257	BC8 10-76 Dispatched E21, L24, TL7, Rescue 1 BC7, BC9, BC10, Division 3 E5 (Command Post Company), E3 & High Rise Unit, Field Communication Unit, Mask Service Unit
2258	BC8 2nd Alarm Dispatched E26, E23, E1, E16, L16, L21, Rescue 4 BC6, Car 17A (Safety Operating Battalion, Division 1 E9 & Satellite 1)
2300	Class E Alarm from ADT for East Tower
2300	Special Call TL14, Tactical Unit 1
2306	Special Call Car 13D (Rescue Liaison Officer)
2321	3rd Alarm Dispatched E39, E34, E40, E22, TL35 BC4, BC12 E207 (Maxi Water Unit), E284 & Satellite 3 Mobile Medical Unit
2330	Special Call 2 Tower Ladders Dispatched TL12, TL13

TIME	ACTIVITY
2337	4th Alarm Dispatched E18, E74, E33, E28, L3
0009	Special Call BC45
0024	5th Alarm Dispatched E14, E53, E24, E55, L25
0052	Special Call BC2
0058	Special Call L8
0059	Special Call L43
0100	Special Call BC1
0130	Special Call Rescue 3, Tactical Unit 2 (for shoring)
0141	Special Call 6 engines and 6 trucks Dispatched E320, E285, E96, E66, E226, E220, TL146, TL114, L19, L47, TL135, TL164
0146	Special Call BC13, BC51
0223	By order of Acting Fire Commissioner Feehan Incident is an "Administrative 8th Alarm"
0430	Special Call 3 engines, 3 trucks & 2 chiefs for relief Dispatched E6, E15, E91, L26, L128/20, L30, BC40, BC49
0445	Special Call Hazmat Company 1 for asbestos evaluation
0602	Special Call TL15 with DeCon Unit (Portable Showers)

APPENDIX C

Photographs

Photo shows the accumulation of combustible wiring that was in the ceiling plenum on the sixth floor at the edge of the fire area.

Photo by J. Gordon Routley

Appendix C (continued)

Photo by J. Gordon Routley

View of the ceiling plenum at the fire division wall shows the accumulation of electrical and communication wiring that penetrates the interior fire division wall above the doorway on the sixth floor.

Appendix C (continued)

Photo by J. Gordon Routley

Corridor on the sixth floor where Engine 65 attempted to advance a 2 1/2 inch attack line.

Appendix C (continued)

Area on the sixth floor showing penetration into the elevator shafts caused by the high pressure master streams.

Photo by J. Gordon Routley

Appendix C (continued)

Photo by J. Gordon Routley

On the sixth floor, this is a view of the structural steel from the underside showing where the sprayed on asbestos fireproofing material remained attached to the steel in an area that was not exposed to water streams.

Appendix C (continued)

Photo by J. Gordon Routley

On the sixth floor, this view shows that virtually all the asbestos fireproofing was washed off the steel that was exposed to water streams. Photo also shows the deflection of the beams and floor deck above due to heat.

Appendix C (continued)

Photo by J. Gordon Routley

The main fire area on the sixth floor looking toward the windows on the 48th Street side of the building. Note the degree of fire damage to contents and exposure of the structural steel above the ceiling.

Appendix C (continued)

Photo by J. Gordon Routley

Investigators working in the area of fire origin wearing protective clothing due to asbestos contamination.